SNAEFELL

MOUNTAIN RAILWAY 1895-1995

Barry Edwards

MIDLAND PUBLISHING LIMITED

Published by
Midland Publishing Limited
24 The Hollow, Earl Shilton
Leicester, LE9 7NA, England
Tel: 01455 847 256
Fax: 01455 841 805

ISBN 1 85780 031 1

Printed in England by
Coventry Printers
Canley, Coventry, CV4 8AW

Designed by
Midland Publishing and
Stephen Thompson Associates

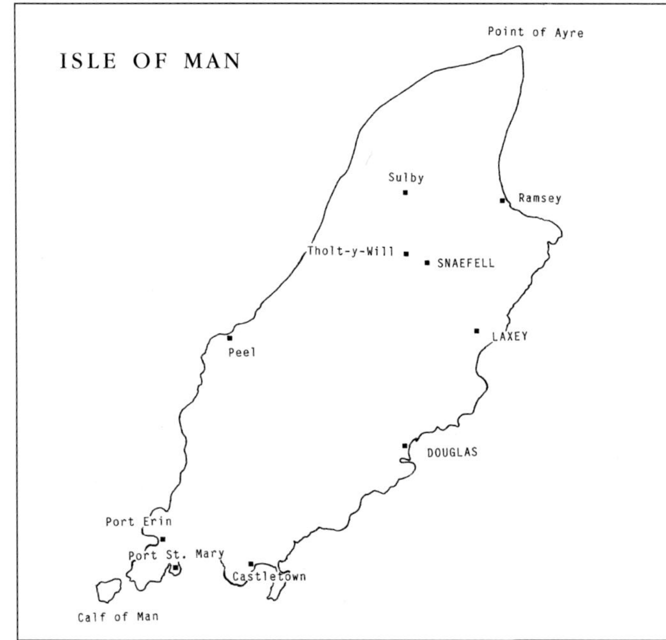

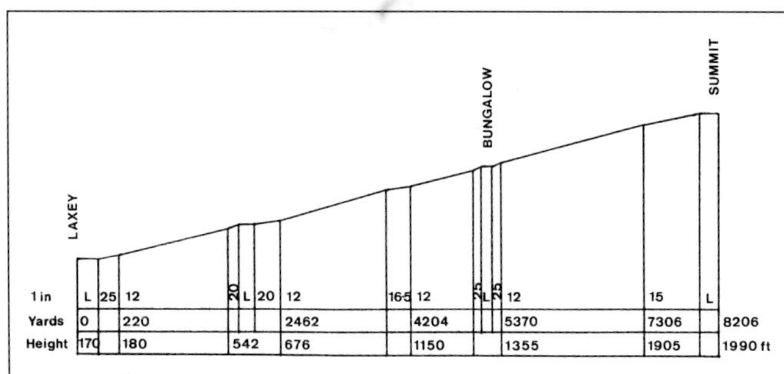

A gradient profile of the
original line.

Title page: **The rebuilt car 5
descends into Laxey from the
summit over the track laid in
1897. The new body for this
car was built by H D Kinnin of
Ramsey incorporating bus style
windows. The clerestory roof of
the original was not reproduced.**

Right: **Map of the
Snaefell Mountain Railway
showing key surrounding
landmarks.**

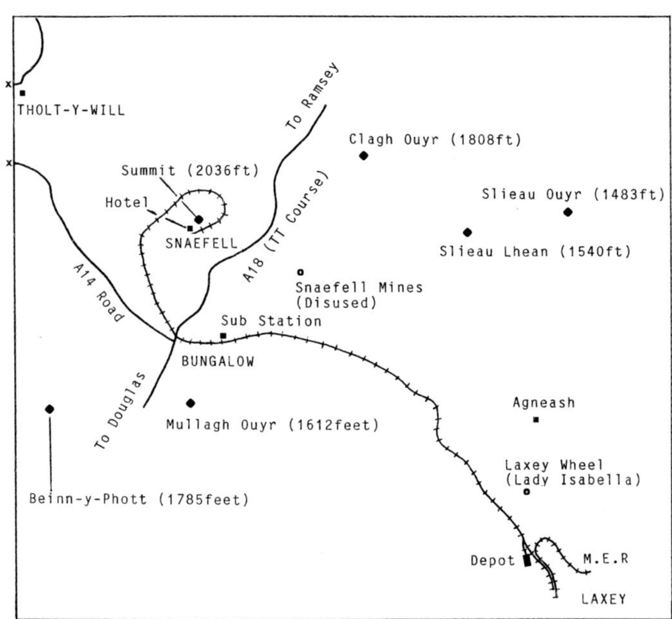

The Isle of Man, which has an area of just 227 square miles, lies in the Irish Sea almost midway between Cumbria and Ireland. The northern end of the Island is flat whilst the middle and south are mountainous. There are 24 peaks over 1000 feet of which only one, *Snaefell*, exceeds 2000 feet at 2036.

with an International Railway Festival on the Island.

In preparing this book I have enjoyed the assistance of many others and thanks are due to Doug Clayton, Peter Johnson, Gordon Kniveton, Lyndon Rowe and W J Wyse for pictorial contributions. Photographs not credited are the work of the author.

Tynwald, the Island's government, has a Department of Tourism and Transport which is responsible for three of the five railway systems, the Manx Electric, the Steam Railway and the Snaefell Mountain Railway. Douglas Corporation operates the horse trams and the Isle of Man Steam Railway Supporters Association the Groudle Glen Railway.

This book is published to mark the centenary of the Snaefell Mountain Railway which is being celebrated

Special mention must go to Keith Pearson for his willing help in providing advice, early illustrations and the drawings. Thanks also to the management and staff of the Isle of Man Railways for their continued support; to my father whose assistance has once again been invaluable, and to my wife Carol for allowing me the time off to compile this work.

Barry Edwards
Ickenham, March 1995

An unidentified car stands at the summit station sometime in 1896, after the car had been fitted with sliding glazed windows but before the addition of the clerestories. The station building measured 50ft by 13ft. The hotel was off the left hand side of the picture. The actual summit of the mountain is above the right hand end of the car. The clock had gone by 1900.
Manx National Heritage

This picture, sold as a post-card inscribed 'Sold exclusively by the guide on the summit of Snaefell mountain, Isle of Man, (2034 feet above the sea)', is dated 1907.
Isle of Man Railways

In 1906 the wooden station building was replaced by this elaborate hotel, which was all but destroyed by fire on the night of 5th/6th August 1982. Remarkably, two road-borne fire fighting vehicles made it to the summit, additional water being brought up by tram.
Courtesy Gordon Kniveton

THE SNAEFELL MOUNTAIN RAILWAY

1895 - 1995

Snaefell, the Isle of Man's highest mountain, got its name from the Scandinavian Vikings, probably sometime before AD300. Translated, the name means 'Snow Mountain'.

The idea of a railway to the summit of the mountain first surfaced in 1888, when Tynwald was asked to approve a plan submitted for a steam operated Douglas, Laxey and Snaefell Railway. Mr J B Fell, the inventor of the 'Fell Incline Railway System', was behind the project but nothing came of it.

The arrival in Laxey of the Manx Electric Railway in 1894 renewed interest in Mr Fell's ideas. It was at the opening speech of a Laxey village church bazaar in the latter part of 1894, that Dr Farrell, a Director of the Isle of Man Tramways and Electric Power Co said 'My friends, I am going to let you into a secret. We are going to build a tramway to the top of Snaefell'.

The Snaefell Mountain Railway Association met for the first time on 4th January 1895 and formally announced its intention to build a line from Laxey to near the summit of Snaefell. The syndicate included Mr Fell and several members of the Isle of Man Tramways and Electric Power Co, including Mr Bruce of Dumbells Bank. Discussions were held about motive power and, after consulting Dr Edward Hopkinson, an early expert in electric traction, the association decided on an electric tramway.

The proposed route, the same as that surveyed by Mr Fell in 1888, climbed at a maximum gradient of 1 in 12 with a total ascent of 1820ft and was nearly all on Crown property, which avoided the need for an Act of Tynwald before construction could begin. In order to incorporate the 'Fell' system between the wheels the line would be constructed to 3ft 6in gauge, six inches wider than the Island's standard, although it is also said that the wider gauge provided

The carved wooden board commemorating the re-opening of the hotel in 1984.

Following the 1982 fire the hotel was rebuilt, though on a more modest scale from that of the building it replaced, and reopened for the 1984 season. Car 2 stands outside the building in 1989. The Civil Aviation Authority pylons tower into the mist that so often surrounds the summit. Further restoration of the hotel took place over the 1994/95 winter.

Car 3, unglazed and clearly displaying the 'Snaefell Mountain Tramway' lettering, stands just short of the summit in this 1895 picture.
Courtesy Gordon Kniveton

Far right: **The unusual point at the summit, once one of several on the line, is now the only one left.**

The 1994 season began on the 30th April. This picture, taken shortly after the arrival of car 6, shows car 2 setting off for Laxey. The Fell centre rail is not laid on the level track at the summit.

greater stability in high winds. This of course is true but it seems an unlikely reason for choosing the wider gauge, as the coastal line was only 3ft and would have to withstand similar conditions on some of the exposed stretches.

Manx Northern Railway locomotive *Caledonia*, built by Dübbs & Co in 1885, was borrowed to assist with construction trains, necessitating the laying of a temporary 3ft gauge rail. *Caledonia* was conveyed from Ramsey to Laxey by ship and then manoeuvred up the Glen road on lengths of track, the piece behind being taken up and laid in front as the locomotive moved forward.

Construction of the line started in January 1895, with bad weather during February delaying the project by about a month, but by early August completion was near. Six trams were delivered from G F Milnes, while Mather and Platt supplied the overhead which was erected on the entire length of the line in just ten days. The wire, 16ft above rail level and suspended from poles situated between the tracks at 110ft intervals, was lifted clear of the bow collectors at each pole, the power being collected by the bow at the other end of the car. As a general rule the wire is removed

Car 2 climbs the mountain past the 55 acre Sulby Reservoir which was opened in July 1983 by HRH Princess Alexandra. The reservoir holds up to 1000 million gallons of water and Druidale Farm nestles near the edge. The Aachen and Westinghouse resistors can be clearly seen on the roof of the car.

Drilling the Fell rail mounting holes *in situ* on the side of the mountain. The flexible drive drill and the clamping mechanism are also of interest.
Isle of Man Railways

The western slopes of *Snaefell* are rugged and in places steep. Car 2 trundles down towards Bungalow on 30th April 1994.

from the upper section every winter to avoid damage.

A power station was built just to the north of the line at 1130 feet above sea level and was equipped with four 6ft diameter 26ft long Galloway boilers. The boilers provided steam at 120 lbs per sq inch to five Mather & Platt 120 hp horizontal compound engines, each with a 7ft flywheel and a speed of 150 rpm, coupled to Hopkinson 60 watt dynamos. A 60ft high iron chimney was placed to the south of the line, the flues from the boilers passing under the tracks to reach it. Water for the boilers was pumped from Tholt-y-Will around two miles away. Power from the station was taken to the feeder poles, situated at one mile intervals, by underground cables.

The 4 mile 53 chain double track line, with one intermediate station ('Halfway', later 'Bungalow') was completed in record time. The Fell rail was laid midway between the running rails, with the upper surface just 4in above the running rail surface. However, during construction, it was proved elsewhere that an electric tramcar could climb a 1 in 9 gradient without assistance, so the Fell equipment was never fitted to the cars although the rail was used for friction braking during descent.

The car depot and Laxey station were situated on a ledge cut out of the hillside just above the village, where the depot still stands. The car shed was brick built with a curved iron roof and measured 119ft long, 20ft 6in wide and 18ft high to the top of the roof. Pits were provided for all six cars.

During the first week of August 1895, *Caledonia* took a party from the London & North Western and Lancashire & Yorkshire Railways from Laxey to Halfway, possibly the only passengers to have been steam hauled over this part of the line.

Various tests were carried out during August and on the 16th of that month the line was inspected by the Board of Trade Railway Inspectorate and an electrical engineer. The inspectors were taken up the mountain in one of the cars, stopping at

intervals to look at the standards of workmanship and to make brake tests.

A newspaper stated that the inspectors were pleased with what they saw but their report to the Governor expressed some reservations. The electrical engineer was unhappy with the line voltage of 550 and suggested a reduction to 520. The location of the power station was criticised and the ability of the cars to continue their ascent in the event of the failure and

A detail shot of the calliper brake shoes on one of the London Transport built bogies. The traction motor and drive gear casings are also visible.

As the line approaches Bungalow
it runs above the A14 Bungalow to
Jurby road. Beinn-y-Phott (1785
feet) forms a backdrop for car 6
as it ascends *Snaefell*. The
lengths of rail lying between the
tracks are there to allow the
incorporation of an additional rail,
to create a 3ft gauge line on
which the former Manx Northern
Railway steam locomotive
Caledonia will run, during the
centenary celebrations of 1995.

isolation of one motor was questioned. What the engineer seemed to overlook was that the motors were wired in pairs so if the power had to be isolated from one, both would be lost!

However, permission to open the Snaefell Mountain Tramway, as it was then known, was granted and it was officially opened on 20th August. The first public services ran the following day, with a return fare of 2s 0d (10p). An average of 900 passengers a day were carried with the six cars operating a 10 minute service.

In December 1895 the Association sold the entire line to the Isle of Man Tramways and Electric Power Co for £72,500, £32,500 more than it had cost to build.

The original Summit Hotel, located a little to the south of the terminus, was extended in 1896 and a new hotel, to become known as The Bungalow, was built at Halfway where the line crossed the mountain road, now the A18 and part of the TT racing circuit.

The Isle of Man Tramways and Electric Power Co considered the distance between their coastal line station and the Snaefell station in Laxey to be too great. In 1897 the Snaefell terminus was moved to a point just above where it now crosses the main A2 road through Laxey and in 1898 a further move brought it into the same station as the coastal line, 4 miles 72 chains from the summit.

The Isle of Man Tramways and Electric Power Co had loans totalling some £150,000 from Dumbells Bank, which collapsed in February 1900, forcing the Tramway Company and many other companies into receivership. An offer of £225,000 from the British Electric Traction Co was rejected in 1901, the Manx Electric and Snaefell lines eventually being sold for £250,000 to a Manchester based syndicate, backed by a merchant banker, in 1902. A new organisation known as the Manx Electric Railway Company was incorporated in London in November 1902 and purchased the lines from the Manchester syndicate for £370,000.

The builders' plate above the cab bulkhead door, inside one of the Snaefell cars.

With just a few yards to go before the Bungalow stop, car 2 descends the mountain on 18th July 1985, a month before celebrating its 90th birthday.

In common with the coastal line, much re-equipping was needed on the Snaefell line and changes were made to the electricity generation and distribution.

In 1904 one of the Mather & Platt steam engines was removed from the power station and replaced by an ECC (Electric Construction Co, Wolverhampton) rotary converter from the coastal line power station at

The Bungalow Hotel, opened at Halfway in 1896, was closed and demolished in 1958. This picture shows car 6 standing alongside the hotel and, although undated, was probably taken in the very early 1900s, the car having its roof board but still displaying the 'Snaefell Mountain Tramway' lettering.
Courtesy Gordon Kniveton

The coastal line celebrated its centenary on 7th September 1993, with thousands of enthusiasts flocking to the Island. Those hoping for a trip up the mountain the next morning were to be disappointed. Car 5 stands at Bungalow, while at this time the author was almost unable to stand up in the wind, yet the summit was covered in a layer of mist.

Ballaglass. This was fed by a 7000 volt AC supply from Laxey, resulting in the remaining steam engines only being used occasionally.

The only serious incident to occur on the line was on 14th September 1905, when three cars were descending in convoy. The first stalled, the

A note in Isle of Man Tramways and Electric Power Co documents of 1897 considered the possibility of a line from Bungalow to Tholt-y-Will but nothing came of this. In 1907 however, following the construction of a hotel and refreshment room at Tholt-y-Will, the Manx Electric Railway Co

second stopped without difficulty but the third failed to stop, colliding with the second and pushing it forward into the first. All three cars were damaged and a number of passengers were injured.

The line continued to be a considerable success and the hotel at the summit proved unable to cope with the vast numbers of passengers. In order to overcome this, an elaborate new hotel with a 100ft frontage was built alongside the railway terminus. It was completed in just four months and opened on 10th August 1906.

started a motor charabanc service over this route. This and the mountain railway services ceased at the outbreak of the First World War.

Following the end of hostilities and the clearing of arrears of maintenance, the line reopened on 10th June 1919. The Tholt-y-Will service also restarted but with different vehicles, as the charabancs had been sold. Three Ford model Ts took over in 1926 and these, in turn, gave way to Bedford coaches in 1939.

To enable the transfer of the Snaefell cars to Derby Castle for overhaul,

In 1907 the Isle of Man Tramways and Electric Power Co introduced a motor charabanc service from Bungalow to Tholt-y-Will where an hotel and refreshment room had opened. About to depart from Bungalow, one of the two original charabancs, MN67 or MN68, has only a few passengers for Tholt-y-Will.
National Tramway Museum

On a fine day car 3 arrives at Bungalow from Laxey on 26th June 1984. The Fell rail stops just in front of the car for the road crossing. During TT (Tourist Trophy) racing the cars terminate here; passengers cross the road by a footbridge and join another car to continue their journey to the summit.

the facilities at Laxey being rather cramped, a mixed gauge siding was laid at Laxey station in 1931/32 where the cars could be jacked up and placed on temporary 3ft gauge bogies.

In 1934 agreement was reached with the Isle of Man Electricity Board to buy the electricity required to operate the line. This ended the generation of power at the line's own power station, which was re-equipped with two Hackbridge-Hewittic transformer rectifiers. These were fed from the grid sub station at Laxey. It was 1964 before any further modifications were made to the power supply, with the installation of automatic switchgear supplied by Bertram Thomas Engineering Limited. This

low area to assist with fuel shortages. This traffic was handled by freight car No 7, nicknamed *Maria*, with bogies borrowed from passenger car No 5.

The railway and coach services resumed on 2nd June 1946. The post-war boom in the tourist industry on the Island provided the line with plenty of passengers, each paying a return fare of 5s.0d (25p) and frequent services were operated.

Visitor numbers began to dwindle in the early 1950s but a decision by the Air Ministry to construct a radar station at the summit provided work for the line during the winter of 1950/51. The overhead wire was left *in situ* that winter to enable workmen and equipment to be transported to

Over the years a number of different vehicles operated the service to Tholt-y-Will. Here we see Bedford WLB MN8685, acquired in 1939, at Bungalow on 20th June 1951. It carried 20 passengers and was withdrawn in 1953. The writing just forward of the rear wheel reads 'Speed 50mph' and part of the roof appears to be some kind of fabric. Late J E Cull, courtesy Doug Clayton

An artist's impression of Bungalow with a somewhat daunting *Snaefell* mountain behind, showing three motor charabancs. If the left hand vehicle is an accurate impression of MN475, built in 1914, it is interesting to note that it lacks stepped seating.
Courtesy Gordon Kniveton

removed the need to man the power station. It was now switched in by the crew of the first up car and out again by the crew of the last down car.

The rail and Tholt-y-Will road services were stopped on 20th September 1939, soon after the outbreak of the Second World War but the line saw some traffic after this, carrying peat down to Laxey from the Bunga-

the summit, the Air Ministry agreeing to make good any damage caused to the wire and poles.

As the Ministry would now need the line all year round on a regular basis, an agreement was reached to allow it to operate its own internal combustion vehicles. These were Wickham railcars, the first arriving in 1951 and a second in 1957. These are

Tholt-y-Will Hotel, in the remote Sulby Glen, is seen here with the other Bedford WLB, MN8874 which carried a 20-seater Duple body. It too was withdrawn in 1953 but was converted to a lorry and finally broken up in 1962.
Isle of Man Railways

Just below Bungalow was the power station. The picture, below right, taken in June 1984, shows car 6 descending towards Laxey and about to pass the building which is now an electrical sub station. The old chimney base can be seen on the other side of the line.

Four of the five Mather and Platt 7ft diameter flywheels dominate this 1895 view inside the power station. One imagines that two or three of these wheels racing round at 150 rpm would have been a memorable sight.
Courtesy Gordon Kniveton

In August 1981 car 2 descends past the entrance to the car sheds over the unusual point-work, which has since been remodelled. Both running rails and the Fell rail moved when the points were operated by the levers to the left of the car.

housed in a small shed on the Snaefell depot site and were delivered with calliper brakes already fitted but were subsequently fitted with horizontal wheels running on the Fell rail, to stop them blowing over in high winds on the upper slopes of the mountain.

Passenger numbers continued to fall and in 1953 the coach service to Tholt-y-Will was withdrawn. The continuing decline in tourism to the Island was being felt all over and, at the end of the 1955 season, the Manx Electric Railway Co advised the government that it would be unable to operate the railway after the end of the following year. The two lines were nationalised on the 6th November 1956, the Manx Electric Railway Act being signed on 17th April 1957 and the Manx Electric Railway Board of Tynwald taking control of the lines on 1st June 1957.

The new company sought to improve the prospects of the railway. The Tholt-y-Will coach service was reinstated using two former Douglas Corporation buses but with little patronage; it was extended to Sulby Glen Station (IOMR) in 1958 and abandoned altogether at the end of that year. The Summit Hotel was redecorated in 1958, whilst the Bungalow Hotel was closed and demolished. The new company also decided to adopt a new livery of green and cream, which was applied to cars 2 and 4 but the latter, in September 1963, was the last car of either line to be repainted; thereafter the original livery was reinstated and has remained ever since.

On 14th September 1905, the only major collision the line has witnessed took place. A shuttle service was operating between Laxey and Bungalow because of racing on the Mountain Road. Three cars were descending in convoy when the leading one stalled, the second stopped in time but the third did not. Either the second or third car was No 6, shown here with serious front end damage.

Courtesy Gordon Kniveton

An interesting picture of an
unglazed car standing on the
ascent track, at what is now
the entrance to the depot. The
two far tracks lead to the shed.
Manx National Heritage

The original Laxey terminus
was alongside the car sheds.
The cars were transferred from
the down track to the up track
using a sector plate. Taken
during 1896, this picture shows
an unidentified car with glazed
windows but not yet with a
clerestory roof. The station
building is to the left, and the
car shed to the right, where
indeed it still stands.
Courtesy Gordon Kniveton

The mid-1960s saw the entire line showing signs of age and corrosion, the Fell rail being of particular concern. As no rail of the type used for the Fell had been produced for around 60 years, Steel Mills of Workington made special rollers in order to produce the required rail.

All the cars were now in need of attention and in May 1977 No 1 was fitted with new bogies built by London Transport. The remainder of the fleet was similarly treated over the next couple of years. The rebuilding is dealt with in more detail later on.

Fire struck at the summit on the night of 5th/6th August 1982, gutting the hotel and as a result the line was closed until 9th August. To assist with the reconstruction, a crossover was installed just short of the station to enable the cars to terminate without having to go alongside the building. Two old 1895 bogie frames were fitted with small wagon bodies to assist in the conveyance of materials used in the rebuilding of the hotel, which reopened for the 1984 season.

During the latter part of the 1980s plans were drawn up to enlarge the car sheds at Laxey, to enable more maintenance work to be carried out there without the need to haul the cars to Derby Castle. The plans were implemented over a lengthy period from late 1989, doubling the width of the shed at the inner end only, to provide workshop and storage space. This extension has now been incorporated into a completely new and bigger shed built early in 1995, following the demolition of the old one.

Further renovation work on the Summit Hotel has followed on from the major refurbishment of Laxey station buildings, which was completed in 1994.

One wonders whether Messrs Bruce and Fell and their colleagues ever imagined that their pioneering line of the 1890s would survive to celebrate its centenary, complete with five of the original six cars. That it has is not just a tribute to them but to all those who, over the years, have worked on the line and its rolling stock, providing us with this remarkable piece of railway which we must continue to look after for generations to come.

A 1981 view of the car sheds. The small building on the left houses the Civil Aviation Authority railcars; the main car shed is on the right. The siding holds the Hurst Nelson wagon, the tower wagon and the Wickham railcar. The car shed has recently been replaced.

The depot points, shown set to enter or leave the sheds. This remarkable piece of trackwork has now been replaced.
Late J E Cull, courtesy Doug Clayton

The original car 5 outside the car sheds in the early 1950s, with the Hurst Nelson wagon. This car was burned down to the chassis in 1970, the end nearest the camera being the only part of the bodywork to survive the flames. Doug Clayton

Works car 7 *Maria* is shunted at the depot by passenger car 3 in September 1954. This was the last appearance of car 7 in service. The bow collector hoops have already been removed, presumably along with the control equipment, in readiness for the removal of the bogies. Doug Clayton

The sad remains of car 7 behind the car sheds in May 1994. The plate sides of some 1895 bogie frames lie in front of car 7.

PASSENGER ROLLING STOCK

The Isle of Man Tramways and Electric Power Co had by early 1895 ordered and received nine tramcars, all from G F Milnes of Birkenhead and in that year a further four were ordered. With several directors of the Snaefell Mountain Railway Association also serving with the Coastal Company, it was not surprising that they turned to Milnes to provide the cars for the Snaefell line.

Six vestibuled saloon cars with ash frames, teak panelling and pitch pine interiors were ordered, each 35ft 7in (10.85m) long, 7ft 3in (2.20m) wide and 10ft 4in (3.16m) high, with 46 wooden traverse seats. Standing room was provided for a further 12 people, although the cars were generally considered full when all the seats had been taken. The windows were unglazed, except in the vestibule, with bars fitted along the full length of the car; a roller blind was fitted to each of the six side openings. The cars were lettered 'Snaefell Mountain Tramway' and carried a red, white and varnished teak livery. They had plain arched roofs and current collection was by two Hopkinson bow collectors, mounted vertically and almost rigidly.

Each car had a Mather & Platt series-parallel controller at the Summit end and a series only at the Laxey end, controlling four 25 hp motors wired in series pairs. Regenerative braking had been considered but was rejected as being too complicated because it required too much driving skill. Braking therefore relied on two mechanical systems, a hand wheel at each end working only on the nearest bogie wheelsets and the Fell calliper brake fitted at the outer ends of the bogies. The inner ends of the bogies were fitted with horizontal guiding wheels running on the Fell rail. The bogies themselves were of a special 6ft 10in wheelbase design, built by Milnes.

Soon after services commenced the cars were fitted with sliding glazed windows, all cars being so treated by April 1896. During the winter of 1896/7 clerestories were added to the roofs, to provide extra ventilation and the seating capacity was increased to 48 about this time by adding a single seat at each end backing onto the cab bulkhead. In 1900 each car was fitted

Inside the car sheds in August 1981, cars 1 and 4 await their next duties. The very cramped conditions inside the shed were relieved by the addition of an extension to the left of car 1, which has now been incorporated into a brand new shed, following the demolition of the original one early in 1995.

with a huge advertisement board mounted centrally between the bow collectors. The boards displayed the car number in the centre top and then the words: 'THIS CAR GOES TO THE SUMMIT OF SNAEFELL – FINE VIEW OF BIG WHEEL – HOTEL AND REFRESHMENT ROOMS ON SUMMIT OF MOUNTAIN'.

During the winter of 1903/04 the summit end controllers were replaced using General Electric (USA) K11 equipment. The Laxey end controllers remained unchanged as they

The car shed being only big enough to house the six passenger cars, works car 7 had to stay outside. It is seen here standing on barrels alongside the tower and Hurst Nelson wagons. Late J E Cull, courtesy Doug Clayton

The first Wickham railcar was purchased by the Manx Electric Company for use as a works vehicle on the mountain line in 1977 and is seen here at the car sheds.

were only used when starting from the Summit and Bungalow or when shunting. In 1906 two stiffening channels were added to the lower faces of the underframes.

The Snaefell cars operated in this condition for many years, the speed on the downward journey being controlled by the Fell calliper brakes. The brake shoes were made from malleable cast iron and padded with crown iron. These materials were chosen to reduce the wear on the Fell rail, obviously with success, as it was to be 1968 before any concentrated replacement programme was implemented.

The controllers were modified again in 1954 to those of the K12 type, then remained unchanged until the late 1970s. The Laxey end controllers had still not been modified at all.

Disaster struck on the 16th August 1970 when, shortly after arrival at the summit on a staff duty, car 5 caught fire. High winds fanned the flames and the inability of the staff to lower the bow collectors resulted in the car being destroyed except for the Laxey cab end, the wind playing such tricks that the paint on the number panel was not even scorched. Various causes have been documented ranging from the regular removal of the bogies from car 5 for use under number 7, to simply the car rocking in high winds, causing the insulation on a power cable to become frayed. The rocking theory caused the roof boards to be removed from the other five cars.

Thankfully, but perhaps surprisingly in view of the falling numbers of tourists, the Manx Electric Railway Board announced that the car would be rebuilt, essentially to the 1895 design but without the clerestories and using modern bus style aluminium windows. The new body was built by H D Kinnin of Ramsey, while the MER repaired the chassis and control equipment. The car re-entered service on 8th July 1971.

The Snaefell cars were by now 75 years old and beginning to show signs of wear, as indeed was the track. The main problem with the cars concerned the motors and control equipment; the actual bodies, although old, were in relatively good condition, probably due to the summer only running and the slow speeds of operation.

Whilst attention was being paid to the track and Fell rail, the Board was looking for ways to update the cars, without losing the appearance and the character of these fine vehicles.

The wonderful Wickham logo on the original railcar.
Doug Clayton

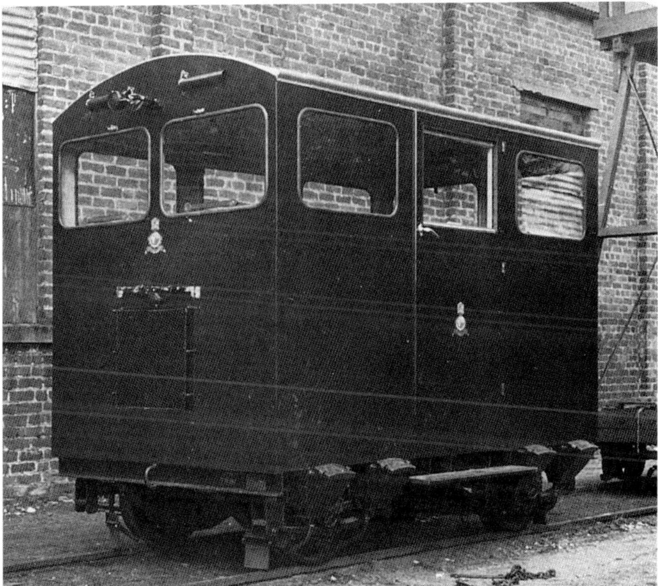

The Air Ministry's first Wickham railcar, purchased in 1951 (works number 5864), stands alongside the Snaefell car sheds. It weighed 2.5 tons and was powered by a Ford V8 engine. Doug Clayton

A 1984 view of the tower wagon showing the Milnes plate frame bogie sides incorporated into the chassis.

An official photograph of one of the original car bogies. The Fell calliper brake shoes are clearly visible. The bogie has a very sturdy appearance, the rule shows it to stand about 2ft 8in high. Mather and Platt

A floral shot of cars 1 and 2 at the depot in 1951. It is interesting to note the differing livery styles and lettering on the massive roof boards. Late J E Cull, courtesy Doug Clayton

Car 2 is jacked up in the shed at Laxey and receives one of its new bogies with roller bearings and traction motors from a former Aachen tram. The Fell calliper equipment was transferred from the original bogies. Isle of Man Railways

In 1973 Tynwald employed Transmark (a subsidiary of British Rail) as consultants to recommend a way forward for both lines. Transmark's report was responsible for the closure of the Laxey – Ramsey section of the coastal line in 1975. Uproar ensued, the government changed and the line reopened in 1977.

Transmark had engaged London suggesting rheostatic braking. The existing motors and control equipment were considered unsuitable for rebuilding and to manufacture the small quantity of parts required for the project would be too expensive. London Transport therefore recommended that secondhand equipment should be used and offered to find some. The cost of the equipment and

In 1897 the terminus alongside the depot was abandoned and a new terminus was built just to the west of what is now the main road through Laxey. This picture shows work under way on the extension; the building to the right is now Brown's cafe. *J H Price collection*

Transport to look at the technical requirements of the lines, in particular the Snaefell cars. The recommendation was that the cars should be re-equipped, possibly with rheostatic braking.

The end of the 1975 season saw the Manx Electric Railway Board make a direct approach to London Transport, who carried out a detailed investigation into the possibilities, again the rebuilding was estimated at £150,000.

In order to justify the financial outlay, London Transport suggested the following points should be taken into account:

1. The projected reduction of maintenance, repairs and labour costs.

2. The cost of rewinding the 1895 motors, by now £600 a time, would be eliminated.

Car 2 stands at the 1897 terminus. The point is similar to that at the summit but was destined to last only a year or so, before the line was extended again. The spoil heap is from the local lead mines. The poles above the 'Snaefell Mountain Railway Station' sign to the left of the picture indicate the route to the car sheds.
Manx National Heritage

Bottom left: When the second extension of the line was completed, it involved crossing the main road through the village, this being done with a single track. Another unusual piece of pointwork was installed where the two tracks went into one. This has now been replaced by a conventional point. The two tracks to the right are the 3ft gauge coastal line.
Doug Clayton

In order to replace the Hurst Nelson wagon, the railway made use of 1895 passenger car bogies made redundant during the re-equipping of the cars. The 'new' wagon (No 1) stands at Laxey with a water tank in 1984.

3. Modern motors would reduce the current consumption of the line by 30%, a saving of £700 per annum at 1976 prices.

4. Modern motors would also reduce the running time and therefore allow each car to complete more return trips on busy days, generating extra income.

5. Rheostatic braking would reduce the wear on the Fell brake shoes, which could last up to six times longer. The pads would therefore not need to be changed every few trips which would increase availability of

Transport or indeed to the Manx Electric Railway Board as this would have been far too expensive and probably unjustifiable in view of the good condition of the car bodies as mentioned earlier, albeit now another few years older.

The search for suitable secondhand equipment got under way, with London Transport contacting various tramway systems, mainly in Switzerland and Germany. Among the requirements of the Snaefell cars was half voltage motors (two in series equal to line voltage).

A view of the 1897 terminus from up the line with two cars in view. The spoil heap and chimney to the left have long since disappeared, the whole area behind the station being changed when the coastal line was extended to Ramsey.
Courtesy Gordon Kniveton

the cars during the working day.

6. The possibility of one man operation.

Tynwald accepted the proposals and a budget of £150,000 was agreed, although London Transport stressed that the costs could not be finally determined until suitable secondhand equipment had been found.

London Transport received a formal contract from the Manx Electric Railway Board during 1976. It should perhaps be emphasised that the complete replacement of the cars was never an option open to London

Several offers were received, the most suitable coming from the recently closed system in Aachen, which had eleven metre-gauge cars numbered 1001-1011, built in 1956/57 by Waggonfabrik Talbot of Aachen, with Kiepe electrical equipment.

The Chairman and General Manager of Manx Electric travelled to Germany along with two London Transport engineers in June 1976 and decided that the cars were suitable. The Aachen authorities had intended to sell all eleven cars to one buyer but

Crossing the road into the present station is the rebuilt car 5, photographed in August 1981 when it was just over 10 years old.

This 1951 picture shows car 1 and the 1895 Hurst Nelson wagon in the approach to Laxey station. The wool shop has given way to some pleasant gardens providing a good view of cars on both lines entering and leaving the station. Late J E Cull, courtesy Doug Clayton

eventually agreed to the sale of just seven at DM10,000 each (approx £2,600 at the time, nearer £4,200 in early 1995). This gave one set of equipment for each Snaefell car and one spare set.

The original idea was to ship just the bogies and other equipment, and sell the bodies to a German scrap dealer. However, the Aachen authorities needed to clear the depot and quoted an unrealistic price for this, so it was agreed to ship the seven cars complete, six to Lots Road power station in Chelsea and the seventh to Douglas, Derby Castle depot.

During October and November 1976, cars 1003/04/05/08/09/11 were delivered to Lots Road via Felixstowe and 1010 travelled via Hull and Liverpool to the Isle of Man. For the record, disposal of the other four cars was as follows: 1001 dismantled, 1002/07 to Wehmingen and 1006 to Viernheim in Germany. The bogies from Snaefell No 1 were dispatched to London Transport's Acton works for a start on the rebuild.

Acton works quickly discovered that to re-use the 1895 bogie frames was impractical because of the unusual mounting requirements of the Aachen motors and their drive system. On top of this, when the Snaefell bogies were cleaned back to metal, signs of cracking were evident. Whilst the bogies were not beyond repair, it was felt that the cost of modification would exceed the cost of a new build designed around the Aachen equipment, with the Fell wheels and brakes being transferred to the new bogies.

Acton started work on building the first of six new pairs of bogies, visually similar to the original 1895 design.

The first pair retained the 1895 axle bearings but subsequent bogies employed new roller bearings, which were fitted to car No 1 at a later date.

Whilst the new bogies were being built in London, the Manx Electric staff at Derby Castle were busy installing the Aachen control gear into the body of car No 1 and at the same time completely rewiring it. The Kiepe control equipment is electro-pneumatic using 24 volt DC control circuits, with the air coming from a compressor and governor; a battery and motor generator supply the 24 volts. The resistors (part of the rheostatic braking requirement) were mounted on the roof of the cars to afford maximum ventilation. The air compressor and governor were fitted under the car, with the rest of the equipment under various seats and in the cabs.

The new controllers have 21 notches, only the first eleven of which are used on Snaefell. Notches 1-9 are series with the resistors in circuit, notch 10 is full series and 11 full series with a weak field. The unused 12-19 are parallel with resistance, 20 full parallel and 21 full parallel with weak field. Braking notch 1 starts the build-up to rheostatic braking and 2-12 give rheostatic braking with resistance in the circuit. 15 mph is averaged at about notch 5, 10 mph at 7 and 5 mph at 9. The rheostatic braking effort does not actually bring the car to a stop, this is achieved using either the Fell or wheel brake.

The work on the car body was complete by Easter 1977, the first pair of bogies arriving soon after, with tests starting in May of that year. The main purpose of these tests was to assess

The Laxey station buildings were completely refurbished in 1994. This view of car 6 at Laxey was taken from within the station building in 1981. The Isle of Man Railways boards have since been removed from the cars.

SNAEFELL MOUNTAIN RAILWAY

When the Snaefell cars need major attention they are put on to 3ft gauge bogies and hauled to the coastal line's Derby Castle Depot at the north end of Douglas promenade. On some occasions the journey is broken at Laxey car sheds where this picture shows car 5 being shunted by coastal line car 21. Late J E Cull, courtesy Doug Clayton

the capability of the resistors when running down the mountain where, unlike urban lines, the rheostatic braking would be in constant use. These tests proved vital as the manufacturer's recommended maximum operating temperature of the resistors was reached in 10 minutes. This meant that to avoid overheating the calliper brake had to be used for parts of the descent to allow the resistors to cool down.

The temperature of the motors when climbing, with all motors in series and a running speed of around 15 mph, was also monitored and found to be well within the suggested tolerance. Running in parallel was also tried experimentally and, coupled with the fact that the new motors were each 61 hp against the original 25 hp, the overall journey time was cut by half. However, such speeds were ruled out as unsuitable for a sightseeing line, quite apart from the wear and tear which they would have inflicted on tramcars, track and overhead wire.

Car No 1 entered service in June 1977. With no major problems coming to light, the decision was taken to proceed with the conversion programme. Cars 2 and 3 were dealt with that winter, with 4, 5 and 6 following during the winter of 1978/79. The overheating of the resistors was tackled and car No 6 was fitted with increased resistor capacity, the equipment being supplied by the Westinghouse Brake and Signal Co Limited of Chippenham, Wiltshire. Tests indicated that this solved the problem and a further five sets of the equipment were ordered and fitted, the final car, No 1, being dealt with in early 1982.

This allowed rheostatic braking to be used throughout the descent. The electric alarm gongs from the Aachen cars were also fitted.

Following completion of the programme and the removal of any useful parts from the Aachen cars the remains of those at Lots Road were broken up. The body of 1010 at Derby Castle was used as a mess hut until 1985 when it too was dismantled.

The cars have since been re-wheeled as it was found that the finer profile of the Aachen wheels caused problems on the Snaefell track with a

The London Transport works plate on one of the new bogies.

number of minor derailments being reported. One such incident when the author was travelling was solved by the crew, without further assistance.

The cars continue to give good service, although the bodies are now beginning to need some attention which is being given by the staff at Derby Castle.

The Manx Electric Railway Board decided to apply a new livery to the cars on both the coastal and Snaefell lines, shortly after taking over in 1957. Only cars 2 and 4 from the Snaefell fleet were treated to this green and white livery. Car 4 is seen here being shunted into Derby Castle depot by a coastal 19-22 series car.
Late A R Connell, courtesy F K Pearson

Flanked by coastal line trailer 57 and illuminated motor car 9, the side displays on which show '1895 Snaefell Mountain Railway Centenary 1995', Snaefell car 4 undergoes a major overhaul at Derby Castle in May 1994.

Below right: **The refurbishment of the Snaefell cars was undertaken by London Transport using equipment from seven secondhand trams from Aachen in Germany. One of these cars (1004) waits on its road loader outside Lots Road power station in Chelsea after its journey from Germany via Felixstowe.** Lyndon Rowe

The interior of one of the Aachen cars after arrival at Lots Road. The Talbot and Kiepe builders plates are clearly visible. Author's collection

OTHER ROLLING STOCK

In addition to the six passenger cars the Snaefell line originally had just three other pieces of rolling stock. These consisted of freight car No 7 – nicknamed *Maria*, a 4-wheeled wagon and a tower wagon.

The origins of freight car 7 are unclear, although it seems certain that the frames at least came from Milnes, and the bodywork possibly from the same source though it was erected locally. It consisted of an open wagon of 6-ton capacity with a full cab at each end but in its own right unpowered. Bogies and control equipment were borrowed from one of the passenger cars, usually No 5, when required. It measured 27ft 8in (8.43m) long, 6ft 6in (2.00m) wide and 9ft 9in (2.97m) high over cabs.

During each winter No 7 carried coal up to the power station for the following season's generating. This continued until 1934 when the generating stopped. From then on it was only used for construction work. It last saw service during the winter of 1954/55, assisting with the Air Ministry developments at the summit. Latterly the freight car spent some time standing on barrels alongside the depot building at Laxey, before being dumped out of sight behind the old station building area.

It is pleasing to record, however, that a reconstruction of the bodywork was completed in 1994/95, with a set of dedicated bogies being made up using bits of some of the 1895 passenger car bogies recovered during the rebuilding of these vehicles. This will allow *Maria* to remain in service along with the six passenger cars and for the first time have its own traction equipment.

The 4-wheeled wagon came from Hurst Nelson of Motherwell in 1895 and was 10ft 2in (3.10m) long and 6ft (1.83m) wide. It was fitted with a handbrake operating on all four wheels. Used to convey supplies to the Summit Hotel, the outdoor life took its toll and it was finally scrapped in 1982.

An overhead tower wagon completed the trio, the origin of which is unknown. This vehicle still exists and is mounted on Milnes plate frame bogie sides, indicating that it was probably built locally, possibly for the construction of the line.

The demise of the Hurst Nelson wagon led to a replacement being built locally, using one of the passenger car bogies removed during their rebuilding. A second similar wagon, again on a former passenger car bogie, was built to assist with the traffic for the reconstruction of the Summit Hotel after the 1982 fire.

The final piece of Snaefell line rolling stock is a Wickham railcar, bought from the Civil Aviation Authority in 1977 for works duties. This is also now dumped at the Laxey car sheds.

SNAEFELL MOUNTAIN RAILWAY FLEET LIST

Car	Built by / Year	Bogies by	Motors	Car type	No. of Seats	Length	Width	Height	Notes
1	G F Milnes 1895	Milnes Special	4x25 hp	Vestibuled Saloon	46	35ft 7in 10.85m	7ft 3in 2.20m	10ft 4in 3.16m	New bogies and control equipment 1977
2	G F Milnes 1895	Milnes Special	4x25 hp	Vestibuled Saloon	46	35ft 7in 10.85m	7ft 3in 2.20m	10ft 4in 3.16m	New bogies and control equipment 1977/78
3	G F Milnes 1895	Milnes Special	4x25 hp	Vestibuled Saloon	46	35ft 7in 10.85m	7ft 3in 2.20m	10ft 4in 3.16m	New bogies and control equipment 1977/78
4	G F Milnes 1895	Milnes Special	4x25 hp	Vestibuled Saloon	46	35ft 7in 10.85m	7ft 3in 2.20m	10ft 4in 3.16m	New bogies and control equipment 1978/79
5	G F Milnes 1895	Milnes Special	4x25 hp	Vestibuled Saloon	46	35ft 7in 10.85m	7ft 3in 2.20m	10ft 4in 3.16m	Destroyed by fire at the summit 16/8/1970
5	H D Kinnan (Ramsey) 1971	Milnes Special	4x25 hp	Vestibuled Saloon	48	35ft 7in 10.85m	7ft 3in 2.20m	10ft 4in 3.16m	Replacement on original chassis. New bogies and control equipment 1978/79.
6	G F Milnes 1895	Milnes Special	4x25 hp	Vestibuled Saloon	46	35ft 7in 10.85m	7ft 3in 2.20m	10ft 4in 3.16m	New bogies and control equipment 1978/79
7	G F Milnes 1896	–	–	Freight car	None	27ft 8in 8.43m	6ft 6in 2.00m	9ft 9in 2.97m	Unpowered, bogies borrowed from No 5.

Passenger Cars: As built these cars had unglazed windows; glazed sliding windows were fitted by April 1896.
The lack of ventilation, when all windows were closed in sunny but windy weather, resulted in clerestory windows being fitted during the winter of 1896/97. Seating capacity was increased in all cars to 48 by adding 2 single seats, one at each end backing onto the bulkhead around 1896/97. Between 1977 and 1979, all six cars were fitted with new bogies built by London Transport at their Acton works and with four 61 hp traction motors and control equipment from tramcars bought from Aachen in Germany.

Freight Car: Nicknamed *Maria*, this freight car saw little use after 1934, except during the Second World War when it was used to collect peat from the Bungalow area to relieve the fuel shortages on the Island.

PASSENGER ROAD VEHICLES

Registration number	Date built	Date acquired	Chassis by	Chassis number	Body by	Body number	Body type/seats	Withdrawn
MN 67	1907	1907	Argus				Ch 16	1914
MN 68	1907	1907	Argus				Ch 16	1914
MN 475 (1)	1914	1914	De Dion Bouton				Ch 18	1914
MN 1053	1920	1920	Caledon				Ch 27	June 1926
MN 1054 (2)	1920	1920	Caledon				Ch 27	June 1926
MN 4416 (3)	1926	1926	Ford 'T'				Ch 14	Sept 1939
MN 4417 (3)	1926	1926	Ford 'T'				Ch 14	Sept 1939
MN 4418 (3)	1926	1926	Ford 'T'				Ch 14	Sept 1939
MN 8685	1933	1939	Bedford WLB	109107			C20F	May 1953
MN 8874 (4)	1933	1939	Bedford WLB	109130			C20F	June 1953
CMN 709 (5)	1938	1957	Leyland KPZ1	200208	Park Royal	B5110	B20F	May 1960
DMN 585 (6)	1939	1957	Leyland KPZ1	201269	Park Royal	B5641	B20F	May 1960

(1) Sometimes shown as MN 479.
(2) Sold to local firm Corlett Sons & Cowley as a lorry.
(3) All three vehicles scrapped in April 1942.
(4) Converted to a parcels van in June 1953, withdrawn June 1957, scrapped June 1962.
(5) Formerly Douglas Corporation No.8.
(6) Formerly Douglas Corporation No.14.

No.	Date built	Built by	Works number	Supplied to	Power source	Weight	Notes
1	1951	D. Wickham Ware, Herts	5864	Air Ministry	Ford V8 petrol converted to Diesel in 1965	2.5T	Sold to MER (SMR) 1977. Dumped at Laxey SMR depot
2	1957	D. Wickham Ware, Herts	7642	Civil Aviation Authority	Ford 28hp Diesel	2.5T	Sold to Wickham Rail in 1991 for further use
3	1977	D. Wickham Ware, Herts	10956	Civil Aviation Authority	Perkins 4.2 normally aspirated Diesel	3T	Replaced No 1. In service
4	1991	Wickham Rail Suckley, Worcestershire	11730	Civil Aviation Authority	Perkins 4.2 normally aspirated Diesel	4T	Replaced No 2. In service

A fine line drawing of car 7 *Maria* reproduced to 4mm scale. S Broomfield, courtesy F K Pearson

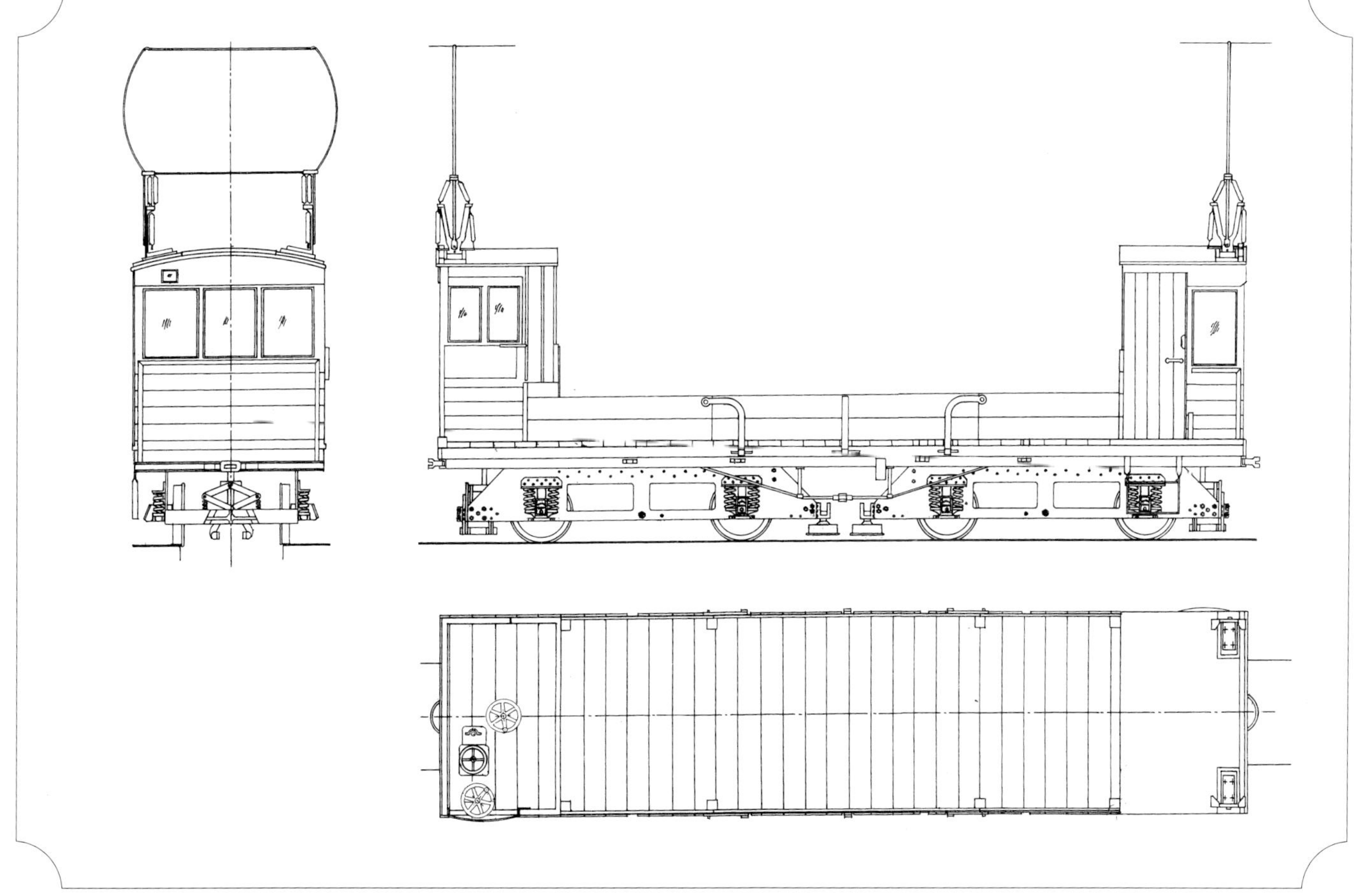

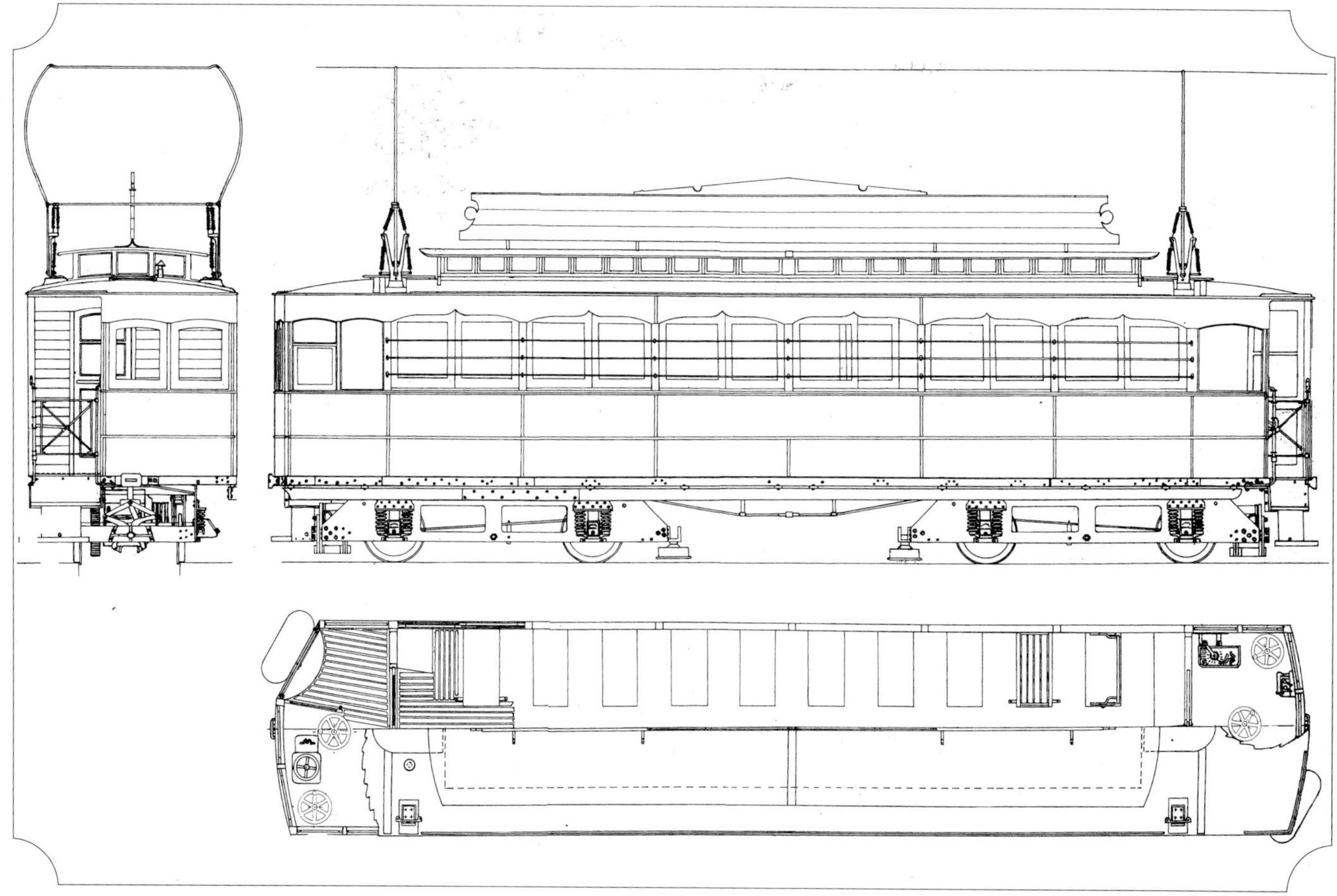

BIBLIOGRAPHY

100 years of the Manx Electric Railway: Keith Pearson; Leading Edge, 1993.
British Trams and Tramways in the 1980s: Peter Johnson; Ian Allan, 1985.
Fleet History of the Isle of Man Department of Tourism and Transport: P S V Circle, 1991.
Isle of Man Album: W J Wise & J Joyce; Ian Allan, 1968.
Manx Electric: Mike Goodwyn; Platform 5 Publishing, 1993.
Isle of Man Railways Fleet List: Barry Edwards; 1994.
Isle of Man Tramways: Keith Pearson; David & Charles, 1970.
The Manx Electric Railway: G Kniveton; Manx Experience & Isle of Man Railways, 1991.
Manx Electric Railway Album: Dr R Preston Hendry & R Powell Hendry; Hillside Publishing, 1978.
Modern Tramway: Light Rail Transit Association; Various issues.
Rails in the Isle of Man - A Colour Celebration: Robert Hendry; Midland Publishing, 1993
The Railways and Tramways of the Isle of Man: Barry Edwards; Oxford Publishing Co., 1993.
Snaefell Mountain Railway: A M Goodwyn; M E R Society, 1987.

4mm scale drawing of a Snaefell passenger car showing the large roof-mounted advertisement boards.
S Broomfield, courtesy
F K Pearson